A Beginners Guide to Consumer Electronics Repair

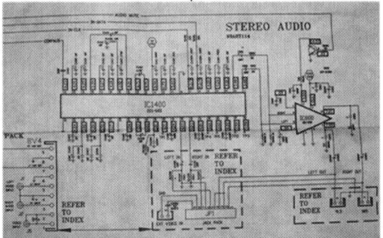

STEREO AUDIO

A Beginners Guide to Consumer Electronics Repair

✦

Hand Book and Tutorial

iUniverse, Inc.
New York Lincoln Shanghai

A Beginners Guide to Consumer Electronics Repair
Hand Book and Tutorial

Copyright © 2006 by Douglas Kinney

iUniverse books may be ordered through booksellers or by contacting:

iUniverse
2021 Pine Lake Road, Suite 100
Lincoln, NE 68512
www.iuniverse.com
1-800-Authors (1-800-288-4677)

ISBN-13: 978-0-595-41171-9 (pbk)
ISBN-13: 978-0-595-85528-5 (ebk)
ISBN-10: 0-595-41171-1 (pbk)
ISBN-10: 0-595-85528-8 (ebk)

Printed in the United States of America

This book is dedicated to a dear friend by the name of Edwin Conrad; whom, passed away in 2002. He helped me to explore the field of electronics, as well as, served as a terrific mentor.

"Formal education will make you a living. Self-education will make you a fortune."

—JimRohn

Contents

Foreword

This book has information that is very condensed but the outline fashion is an effort to keep the topics separated. As a reader, you should find this book useful for an overview of general fundamentals of electronics. The emphasis is on symptom troubleshooting, diagnosis and repair. The information is informative guideline and not lesson type information. It is recommended to take watch for particular types of information, which will be helpful. For instance, math equations, component measurement types and basic function of each component which can be reviewed and practiced to help make your abilities stronger. Another example, circuit tracing to verify the path of current may, also, be helpful.

Preface

I started out wondering, immensely, about which, why and how any particular circuits and/or components were involved in faulty equipment. I asked a lot of questions at first about troubled circuit malfunction and/or about the makeup of particular circuits and functions of components. Thinking through my experiences, I decided to put together quick facts, general synapses and detailed descriptions of the major factors of the knowledge needed to understand electronics and their operations. My main goal for writing this book was to layout the essential questions for all basic tv operation and functions. It includes the circuit descriptions and diagnoses of typical symptoms of tv malfunctions. The layout is to display information for ease of understanding and reference to the reader. Component description and functions and details about tools and testing can be answered right here in this handy little field book and tutorial. This book has each topic with bolded headings and in the table of contents, for fast access and easy learning. Enjoy!!

List of Abbreviations

A (amp) Ampere

AC Alternating current

AFC Automatic frequency control

AGC Automatic gain control

Ai Current gain

AM Amplitude modulation

Ap Power gain

Av Voltage gain

BJT Bipolar junction transistor

BW Bandwidth

C capacitance or capacitor

CRT Cathode ray tube

CT Total capacitance

DC Direct current

EMF Electromotive force

F Frequency

FET Field effect transistor

FM Frequency Modulation

f_r Frequency at resonance

H Henry

I Henry

Hz Hertz

I_B DC Base current

I_C Collector current

IC Integrated circuit

I_e Total emitter current

IF Intermediate frequency

I_R Resistor current

I_S Secondary current

I_T Total current

JFET Junction field effect transistor

L Coil, inductance

LC inductance-capacitance

LED light emitting diode

MOSFET Metal oxide semiconductor field effect transistor

N Number of turns in an inductor

P Power

Q charge or quality

R Resistance

rf Radio frequency

RL Load resistor

UHF Ultra high frequency

UJT Unijunction transistor

V Volt

VA Volt ampere

Vav Voltage average value

VBE DC voltage base to emitter

Vc Capacitive voltage

VCE DC voltage collector to emitter

VHF Very high frequency

Vin Input voltage

VL Inductive frequency

$Vout$ Output voltage

W Watt

XC Capacitive reactance

XL Inductive reactance

Z Impendence

Introduction

I'll start off saying that there is much more to learn about electronics than you will learn in this book; as I am sure you guessed. But, you will learn everything you need to know about repairing consumer electronics with emphasis on televisions; which, are built extremely similar to computer monitors. The types of circuitry you will learn about are relevant to many types of consumer electronics but this book will describe them using a television. You will be able to use this book as a field reference, as well as, for learning the knowledge and the skills.

Your first question may be, how many and what kind of tools will you need. And, how much is it going to cost me? An oscilloscope is the biggest purchase and is a necessary tool for every tv service technician. Oscilloscopes can cost from $300.00 to $2,000.00 depending on size and capabilities. And, then you will need a meter that will measure ohms, volts and amps, screwdrivers, a soldier gun and in many cases a schematic (a diagram of the circuitry of any given tv) of the individual tv set. Those are your essential tools.

This book can help prepare you to take the Associate certified Electronics Technician exam, which, given at the Electronics Technicians Association, International. The exam can be taken by individuals that have less than two years experience or technical education in electronics. After, the first certificate is received; there are many journeyman exam options. To take the journeyman exam one has to have at least two years of experience and/or education in the field of electronics. Some options are Avionics, Biomedical, Computer Service Technician, Radar, Consumer, Industrial, Satellite installer, Network Systems but here are many more.

Need to know information in the major sections, for the Associate certified Electronics Technician exam, is basic math, ac and dc circuits, transistors and semi conductors, electronic components and circuits, instruments, tests and measurements and trouble shooting and analysis. This is a short list of specific topics that will be discussed on the test, capacitors, inductors and transformers, symbols, biasing, diode circuits, thyristors, power amplifiers, diodes in series and parallel, operation amplifiers, networks of circuits, amplifier tests and digital logic. This book explains some of these topics, but, if you are anticipating taking the exam, you will need to study and practice material given from the examination center, or elsewhere.

Safety and Precautions

We will start off with the most important issue, which is, technician safety. The technician is very often subject to electric shock; therefore, prevention and treatment are key points for our initial chapter. Testing for shock hazards can usually be done by using a voltmeter. A common time to check for electric hazards is when an oscilloscope is plugged into a voltmeter and the voltmeter is plugged to the television. An isolation transformer can be used to place between the wall receptacle and the tv to prevent against electric shock, such as, sudden surges; as with, lightening. Also, it protects against a possible ground system that keeps the chassis (the electronic circuit board) hot all the time.

Jewelry should never be worn when working on electrical equipment. Picture tube voltage sources and low input voltage power supply are two common ways to get electrocuted. Picture tubes can be dangerous, so precaution is necessary working in these two areas. It is a very good idea to keep one hand on a non conducting surface to reduce the chance of shock. It is also a good idea to stand on a rubber mat while working.

The second anode can be used to drain the electricity from the tube. The tube should be drained before handling. To do this simply put the tip of a screwdriver underneath the anode and connect the metal part to a ground (chassis); make ground connection first. Implosion of the picture tube is another place to be very careful. The screen portion of the tube should always be towards the front of your body, or, it could explode and fragments would shoot into your body.

1

Essential Functions of Television Components

In the first chapter, we will discuss the essential functions which are the basis of troubleshooting for the majority of problems with televisions. Within the picture tube, the television produces pictures, such as a printer prints on a sheet of paper. The first line is done from left to right and then it starts on the next line, the same way. The electron beam is sent to the screen of the picture tube from the electron gun and the phosphor on the screen glows when the beam hits it. The picture on the screen is created by the frequency sent by the broadcast station. This is done by the electron beams moving at a rate in accordance to the signal. The picture elements are grey dots all over the screen and are produced by the beam turning on and off very fast. Then, the frequency and sweep rate are manipulated to create the picture.

The electron beam is retraced to scan again by a vertical sync pulse and a horizontal sync pulse. The horizontal sync pulse is sent when the beam is to be sent back to the other side of the screen. And, the vertical sync pulse is to send the beam back to the top of the screen, and start again. A typical picture can come through in segments called fields starting with a horizontal sync pulse, then, a line of picture information, then, a horizontal sync pulse. It is done in this way until it reaches the bottom of the screen and needs a vertical sync pulse to get to the top again. A raster occurs when there is no broadcast signal.

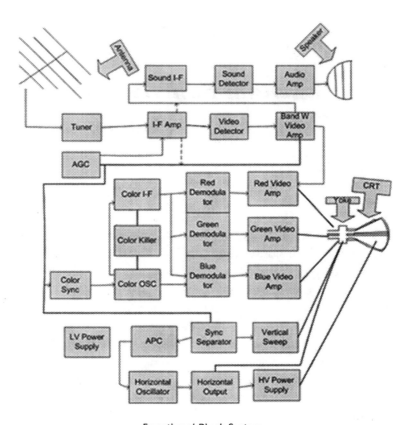

Functional Block System

The functional block system is a good place to start to show how the power inputs make a difference to the operation of the television and the troubleshooting of the television. Each part of the block system is the troubleshooting foundation. The television functions, for troubleshooting purposes, are usually considered from the raster, picture, sync and audio. In this chapter, will be a brief description of the important aspects of each of these sections, followed by, troubleshooting, for the possible malfunctions, that are directly associated with these circuits. To learn these blocks is essential and you should know the interdependence and the functions, the symptoms of possible problems and the troubleshooting procedures for each block.

Producing a raster is done by several different components, power supply, horizontal circuits, high voltage, and vertical circuits. One very basic consistency of the power supply is that, it has different voltage sources following the main power supply (from ac to dc power), automatic phase control, horizontal oscillator, horizontal amplifiers and high voltage. High voltage leads to the picture tube by a small (usually three inches in diameter) rubber ring to the back of the picture tube. So, the raster would disappear if any of these blocks malfunctioned. Each block has different circuits to furnish power and signal to. For instance, the power supply block has many circuits, horizontal output volts +125 video vertical outputs +80 volts, audio circuits +24volts, color circuits +12 volts and all other circuits +12 volts.

The picture is produced by several different components, tuner, I-F circuits, video circuits and the picture tube (CRT). A particular circuit function may be found to be a problem after troubleshooting and testing the brightness, contrast, fine tuning, and other picture changing controls. There are three signals that make the tuner operate, by way of, receiving and converting the signal. The signal is brought from a radio frequency amplifier and goes to the mixer where it mixes with a signal generated by the oscillator circuit, in which, creates the intermediate frequency (if). The oscillator frequency is always 45.75 MHz higher than the picture signal. Many signals are produced; the sum of the two signals, the difference between the two and other signals are created. The i-f signal is always 45.75 MHz.

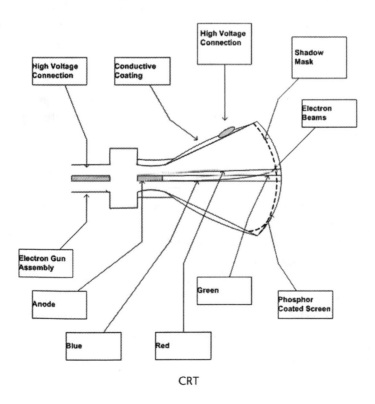

CRT

There could be two or more i-f amplifiers and they are controlled by the automatic gain control circuit (AGC). The (AGC) takes a sample of the incoming signal, turns it into dc and then causes the i-f amplifiers to gain or lose amplification of the signal. This keeps a constant picture. If the i-f signal processor or the AGC circuit is bad it could create loss or degradation of the picture. The video detector is the last circuit which links the i-f circuit to the video circuits. The video circuits now receive the signal and amplify it, by using at least two amplifiers. And, the amplified signal goes directly the picture tube.

The contrast is the way to adjust the amount of video signal; more video means blacker. A color tv has three video inputs, one each, for red, blue and green colors, and a black and white tv has one. There are three phosphors on the screen red, blue and green. The CRT (cathode ray tube) amplified video signal is controlled by the video amplifier. The crt is where the sweep, signal and high voltage must be maintained to produce a picture. Anything before the crt and after the tuner will produce a problem with the picture.

Picture synchronization is done by a few different components within the Sync separator, differentiator, integrator and vertical and horizontal circuits. There are three main problems that can occur because of the synchronization circuits going bad. The loss of both horizontal and vertical sync is created by a faulty separator circuit. Poor or no horizontal sync is caused by the differentiator circuit. Rolling or vertical sync problems are because of this sync circuit. These circuits may be fewer for problems because there is less signal voltage and there are no integrated circuits and transistors that will go faulty. Synchronization is also provided by the horizontal APC and the horizontal oscillators. Both oscillators being faulty would cause no picture.

The sound is produced by a sound i-f amplifier, a sound detector and audio amplifiers. Our first test here would, probably, be the video detector because we can check the quality of all the signals from this point. The sound signal comes from the broadcast station with the video signal and is separated here to produce sound. Because of this, many problems, such as, sync, sound and picture can be created from faulty tuner, video and i-f circuitry. Different tvs are designed with the sound signal being separated after the video and some are taken off before the video at the i-f circuits. The sound i-f circuits secure the sound signal, and then, go to the sound detector that outputs an audio frequency to the audio amplifier; there are two or more sound amplifiers. The sound i-f is 4.5MHz.

2

Tools and Techniques

Aside, from the essential tools, oscilloscopes, multimeter (volt, ohm, and amp meter), screwdrivers, a soldier gun and a schematic there are many other types of equipment that can be used. A schematic can be purchased from the manufacturer of the consumer electronic or by an independent company that sells schematic diagrams. There are signal generators, analyzers, CRT testers and transistor testers, adjustable power supplies and color bar generators.

Many times, we will be unsure of where a particular problem is by simple diagnoses, so then, we can use these tools. Often times a technician may use the approach where he/she will change a whole module if there is one handy and inexpensive; but this is not always the case. Many technicians would say that the oscilloscope is the most useful piece of equipment because it can measure voltage, frequency, the direction of the signal and signal distortion. The voltmeter is a very handy to use as a portable device for measuring voltages.

You will need to know what value of measurements that you should be reading in order to make the proper diagnosis. The resistors have a color code that tells you what the resistance should read for each particular resistor. The transistors have the value stamped on the transistor in numbers. The measurement for the transistor is somewhat universal; silicone transistors should have .7 v and geranium transistors should have .3 v. The capacitor can be tested by allowing current to flow through from an ohm meter and then you can discover if the capacitor is actually the component at fault by measuring the transistor's voltage drop. The amount of voltage and measurements for all devices are on the schematic.

To test a circuit, after you have localized the problem, test the component that would most likely be faulty, check voltages around the circuit and analyze current, within the circuit. There are two types of components active and passive; active components fail most often. Active components are the transistors, diode, integrated circuits (I-C) and amplifiers. This type may be used in horizontal, vertical and audio power supply circuits.

⸱ The inactive components are the transformers, resistors and capacitors. The active components are very much more likely to fail and, especially, if they are high powered. For example, a high voltage drop across a transistor most probably is that the transistor is faulty; however, a preceding capacitor or a resistor could be leaking causing the high voltage drop. So, in this situation you would want to check the resistance across the capacitor and the resistor. Because the least resistance would make more voltage form other parts of the circuitry to get to the transistor. Note, when diagnosing resistors, usually, they increase rather than decrease.

Now, we are ready to replace the parts. When soldering the components on the circuit board there should be absolutely no soldier splashes or sharp points on any other part of the circuit board or components; if this happens they will, respectively, cause a short or an arc. There should be no soldier splashes anywhere on the components. Use exact same size components, and use low wattage irons for transistors, i-cs, silicon-controlled rectifiers and diodes. A word of caution is when using any equipment in creating voltages you should turn off the tv or arcing can occur and cause damage to the components.

Some major problems are easy to find because it is obvious and easy to test. But, some problems are less drastic of malfunctions and are more difficult to locate. This is, because, the components that are bad may be misjudged when tested, causing the technician to overlook very small defects. An example, of this, may be a slight loss of a vertical sync. Quite possibly, a capacitor is leaking, a little, or a resistor has increased in value, a little.

3

Television Assembly, Disassembly and Cautions

One very important rule is to never remove any parts while the power is on. If you remove parts while the power is on, the part could be damaged or other parts in the circuit could be damaged, as well. Reassembling a tv can be quite complex Some technicians prefer to write down notes or sketches. When disassembling the tv, it could help you by taking notes, to keep good track of which type of hardware goes with each particular section of the disassembled tv. This chapter will review disassembly and removal of parts of the tv that may have to be removed. The only tools that you need to disassemble the color television receiver are a nut driver, a screwdriver and pliers.

When beginning to take apart the set the first thing that needs to be done after the removal of the back of the tv is to discharge the tube. The tube can be discharged by placing a screwdriver behind the second anode until it meets the metal prongs and connecting a wire to ground. It is recommended to wait a few minutes and discharge it again, because voltage can build back up in the tube. Now, the second anode lead can be removed. Some new sets have a bleeder circuit. Remember to be careful when working with the crt. The picture tube can implode, especially, at the neck because there it is the thinnest. Also, be careful not to scratch the tube with any metal objects. With some tubes the neck is cemented, and so, the convergence must be done by manufacturer's specifications.

The first step, in removal of crt components, is to disconnect the grey wire that is connected to the DAG ground clip on the top right corner of the crt. Next, we have to remove the circuit board that is mounted on the back of the neck of the crt. The board has to be shimmied out of the neck. Now, find the rubber ring (degaussing coil) and disconnect the wire at the board. Next, the removal of the button board can be done by the removal of a few screws. Next,

7

the main module can be slid out. When removing the purity yolk assembly it is very important to make a mark on the neck of what position it was in, so that, it can be placed back in the same position. This is very critical and can be done with a marker or a note of a measurement. The crt can now be taken off, if needed.

A little later, in this chapter, we will go over the make up of the different sections of the tv. But, first we will need to go through the reassembly procedure of putting the tv back together. The reassembly procedures are the same as reversing the order of disassembly. Check to make sure wires are not being caught next to sharp objects and or that all wires have a tight connection. The set can now be turned on and converged because the purity ring and the yolk has been replaced, as removed. Then, the back can go back on the tv.

We have already, gone over the functions of the particular parts of the tv, now we will discuss the make up of the parts, to describe the significant component structures. There are many of the components that slightly differ from there very similar counter parts. One example, of this, would be the zener diode as compared to a normal diode. The zener diode is designed, so that, it regulates voltage by breaking down and allowing voltage to pass at a certain voltage. The normal diodes are simply for passing voltage in one direction. In a later chapter, the most common devices, such as these, will be described. Tuners are mounted near the audio circuits on the main module. Frequencies through the tuner are 50 MHz up to 1GHz. There are two unique components in the tuner that help it operate at desired levels of voltage. The varactor diode and the RF switching diodes; the varactor acts as a capacitor and changes with varying voltage. The RF switching diodes switch on and off allowing different levels of frequency to come through. However, keeping this in mind may not be needed because most often the whole tuner will need to be changed.

The microprocessor contains analog processing and digital processing circuits. The circuits are counters and are made of flip flop circuits. A pulse generator is what drives the clock and a crystal stabilized oscillator is what drives the pulse generator. The microprocessor is programmed and designed to do many jobs; the inside circuits are often referred to as memory. The impulses are given values by the use of binary numbers and this is how this microprocessor (often called microcontroller) is enabled to count.

A single i-c can house up to four amplification stages. After, the first stage, in the tuner, the i-f is next in the main module. There is test gear to realign the i-f circuit; however, the circuit should never have to be realigned. Tuning of the i-f circuit can be adjusted throughout the whole passband. The i-f signal then goes to the detector. The synchronous detector is used in the modern type of televisions; however, diodes are still in some older sets. The detector works by using transistors that are designed in the circuit to collect signal in phase. Two transistors will collect signal during one half cycle and the next half cycle will be collected by two other transistors. This would, undoubtedly, provide linearity in the signal.

The audio circuits are controlled by decoupling with capacitor to block the signal that would cause error. A phase locked loop, which works very similar to an AFC circuit, used in the tuner, or the ACC, used in the ACC color circuit. The frequency comparator will generate a d-c error signal. The generated frequency is the difference of the two frequencies coming in. The signal is fed back to the oscillator continuously, maintaining the frequency of the incoming F-M 4.5MHz. phase locked loops.

4

Circuit Analysis

One very important law is the ohm's law Resistance also known as ohms=volts (v) /amps or amperes (a). Here is just one example of a common scenario observing the ohms law at work in a circuit. The ohms law can be used to calculate volts, amps and resistance; E=I*R, I=E/R and R=E/I. As a voltage decreases, the resistance, also, decreases. In this case the component could be defective, or another component in the direct circuit is faulty or was activated. Another, very important law in electronics is the Kirchhoff's laws, the law or current and the law of voltage. The Kirchhoff's current law states that the sum of the currents entering a circuit must equal the sum of the current that come out of the same point. The Kirchhoff's voltage law states that the algebraic sum of instantaneous EMF's (electrical motor force) and voltage drops around any closed loop circuit is zero.

The power triangle is also a very popular way to remember how power is distributed in a circuit; P/ (I*V). In the next few sentences, I will give you the equations that can be derived from the comparing power (P), current (I), voltage (V) and resistance (R). Voltage is sometimes referred to as (E) from (electromotive force). But, first I will need to tell you the units of measure for these values, P=Watts, I= Amps, V=Volts and R=Ohms. The equations for power are, P is equal to V*I, V squared /R, and I squared * R. The equations for volts are, V is equal to I*R, the square root of P*R and P/I. The equations for ohms are, R is equal to P/I squared, V/I and V square / P. The equations for amps are, V/R, the square root of P/R and P/V.

Electric circuits are made of three main values, voltage, current and resistance. Voltage is electrical pressure, current is the flow of energy through electrons and ohms is resistance within the circuit. The schematics will have printed on it what measurements should be where. However, it is good to know how the circuits work. Knowing how the electricity flows and how the voltage and current are laid out amongst the circuits. The circuits can usually be broken down into three types, series, parallel and series-parallel.

In a series circuit there is only one path for the current to follow. In a parallel circuit the current is split up into how many parallel paths that there are in the circuit. The parallel circuit can have different values of circuit and semiconductor devices to help the desired distribution of current and voltage. When calculating and measuring the resistance for the series-parallel circuits, simply do the separate calculations for each part of the circuit and then add them together.

The measurement of each circuit is as follows. In a series circuit the value of the resistors will be added up for total resistance. The current in a series circuit is the same for every component. Voltage in the series circuits can be calculated by using ohm's law. Also, the theoretical measurement of each component can be done, simply with ohm's law. But, to get the calculation when there are many resistors, the components get divided by the resistance of each resistor and voltage (V1/V2 = R1/R2). In a series circuit the capacitance is calculated by summing up the reciprocal of each individual capacitances and then taking the sum. The total inductance of non-coupled inductors, in a series circuit, is made by taking the sum of each inductor.

In a parallel circuit, the resistance can be calculated by adding together the reciprocal of each resistance. Then, the resistance is the reciprocal of the sum. The current in the parallel circuit is calculated using Ohm's law on every loop and adding them together. With the parallel circuits, the voltage variable is the same throughout the circuit. In a parallel circuit, the calculations are made by taking the sum of each capacitor. The total inductance of non-coupled parallel circuits is found by taking the reciprocal of the sum and the reciprocals of each of the inductances.

Another, major characteristic of common circuitry is resonance in a circuit. In a LC series circuit the resonance is reached by the capacitance reactance and the inductance reactance canceling each other out. Also, in the RLC series circuit, the LC parallel circuit and the RLC parallel circuit, there are methods of measuring the resonance and other important values. Frequency (fo) or (fr), (Q) charge or quality and (BW) bandwidth are amongst the other important independent values. If one increases and the other decreases, the formula is fo = (BW) * (Q). To further calculate these values the equation can be manipulated to find any particular measurement needed. For instance, fo =1/ (2PI*the square root of LC).

5

Component Descriptions

We will discuss, in this chapter, all circuit components. The first thing that needs to be specified, especially, to the very beginner, is that components in the same category of device have many different types and functions. In this chapter, I will outline the components by category. Each category will start with the most usual components type of that category. There will be some particular devices in electronics that I do not mention, but the essential components to get started with electronics repair are in here.

In the back of this book is an appendix that shows the schematic symbol for many essential components. Short descriptions of some of these components are in later chapters. The schematic symbols vary depending on who is using them. Sometimes, there are several different pictures to symbolize the same device. Also, in the appendix is a color chart to show the resistance of resistors and a color chart to show the capacitance of all capacitors.

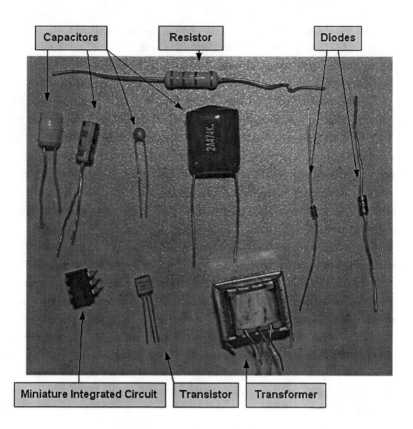

Capacitors   Resistor   Diodes

Miniature Integrated Circuit   Transistor   Transformer

Diodes are semiconductor devices designed to restrict the movement in all but one direction. There are several types of diodes and most of them have a band on the cathode end indicating the positive side. The anode side is where the direction of current is entering. You must have seen the small lights on a piece of electronic equipment indicating that it is turned on. That is a light emitting diode.

Some common diodes are zener diodes and varactor diodes. Typical diodes are used as a rectifier to keep current in one direction and to maintain dc current. Zener diodes are made mostly for voltage regulation because it allows current back through it in the opposite direction, at a specific voltage. Varactor diodes are designed for a capacitance characteristic. They are commonly used in, amplifiers, oscillators, phase-locked loop and synthesizers.

Diodes can be used in conjunction with many types of circuits to achieve specific results. Common circuit types are steering circuits, frequency multiplier or mixer circuits, switching circuits and as mentioned above rectifying and regulating diode circuits. One example, is the result of the function of a couple of diodes, in a circuit, to turn a-c current into d-c current called; a full wave rectifier.

ⁱ **Transistors** are three pronged semiconductor devices that are commonly used for amplification, switching, voltage stabilizing and modulation of signals. This is done by the transistor acting as either a gate or a switch. There are metal transistors where the case is the collector and the emitter and the bases are on the bottom side, which mounts to the board. One prong is called a base, one prong is called a collector and one prong is called an emitter. These prongs are typically represented by the letters C, B, and E. On the transistors, when they are not labeled, check to see the flat side and hold the transistor down with the flat side facing front. In this position, the base will be on the left, the collector in the middle and the emitter on the right.

It was the transistor that made electronics fit into small and portable packages. The most common types of transistors include bipolar junction transistor (BJT), field effect transistor (FET), junction field effect transistor (JFET) and the metal oxide semiconductor field effect transistor (MOSFET). There are two ways that transistors are made in configuration. One is the NPN type and the other is the PNP. The name comes from the fact that the transistor plate that is sandwiched between the other two plates is either doped or has holes created in the plate. The one that is doped is doped with a material that will causes more negative energy (electrons) to flow through and the other causes more protons to flow through it.

The BJT transistor has two junctions. The structure is NPN and PNP as mentioned earlier; as with all transistors. The characteristic that allows them to be bipolar is that they conduct current in both directions using both polarities. The NPN emitter collects electrons and the PNP collects holes. The two junctions are emitter-base and emitter collector. The FET and the MOSFET are both JFET transistors. The FET transistor is very similar except it accepts energy in only one direction or either direction. The JFET is different from the transistor because it acts as a vacuum tube. This is possible, because of its high impedance and, the fact, that it is voltage controlled. MOSFET is designed differently by having a much thinner layer of oxide allowing it to be better used for higher impedance.

Capacitors are devices that store energy and feed it back to the circuit when prompted by the other components. Capacitors can, also, be used to block a-c current from a circuit because it cannot accept d-c current. The capacitor has two prongs. One prong is shorter than the other, indicating negative polarity, and the longer prong indicates positive polarity. There are many types of capacitors but in two main categories, the Electrolytic capacitors and the non-electrolytic capacitors. The electrolytic capacitors are able to store much more energy than the others. Next, I will list common capacitor types and there characteristics. Electrolytic capacitors are generally used for timing circuits, ripple filters and to store charge.

There are several types of electrolytic capacitors that have varying characteristics. Some are very small size, shaped for tight fits and also are made of different materials, making them less vulnerable to explosion or cracking. Some other capacitors are used for filter circuits or resonant circuits. Capacitors come in many different shapes and make up. Adjustable capacitors are color coded and are designed to create variance or trim a circuit current. Capacitors are made of many different materials and each one suited for there own particular jobs. The deciding factors for the engineers to choose a particular capacitor for a job are frequency, voltage, temperature stability, physical size and cost of the capacitor.

Resistors are components in a circuit that opposes current to the specified value of each resistor; controlling the amount of current in a circuit. The resistor is used in a circuit to keep the current the desired value, dissipate heat, reduce voltage, synchronization of current, and regulates the speed of current. The types of resistors are fixed, variable, varistor, thermistor, light sensitive resistors and sensistors and surface mounted resistors. Fixed resistor will be the ones that you will be working with mostly in tvs.

Resistors are radial shaped and all look very similar in shape. They have color bands on them, indicating the value of the resistance and the value of the tolerance. The typical types of band formats are four and band color codes on the radial resistors. The five band resistor is used for bigger resistor values. There is a table in the appendix showing how to read the resistor bands. The first line with the group of line is the 1st digit, the second line is the 2nd digit, the third line is the multiplier and the last band separated from the group is the tolerance. The five bands simply have one more line as a third digit before the multiplier.

Inductors are components that oppose any change within the circuit; inductor and coil are synonymous. Inductors have some similar and opposite characteristics of a capacitor. The inductors capacitive reactance (Xc) will decrease when an applied frequency increases and the inductive reactance (Xl) will increase. They are used in impedance matching circuits (impedance is the equivalent to ac as resistance is to d-c), resonant circuits, analog circuits and signal processing circuits. Inductors oppose voltage while capacitors oppose current. Inductors, often times, are referred to as chokes and are made of either powdered iron core or ferrite core materials. Inductors are made in many different types and sizes, mainly, fixed and variable. Inductors, also, have a color code to match with the color bands on the case.

Transformers transfer energy to a circuit using a magnetic coupling. The different types of transformers are step up and step down, isolating and variable and distribution transformers. In tvs, you will not see the distribution transformer because those are very high voltages. In tvs the transformers are used to step up and to step down voltages. There are two windings of coil that create a magnetic flux; and, if the first coil has less winding, the voltage will be stepped up and vice versa.

Most transformers in tvs are shaped as a rectangle box. There are three primary losses in transformers, which are, eddy currents, hysteresis and copper losses. Because, of these characteristics, transformers typically give a ninety five percent efficiency rate. In the tv there are power transformers and audio transformers. Transformers can only operate using ac power. It cannot accept dc power. Transformers in tvs are very simple to measure. The measurements can be taken with a scope, but not directly on the transformer.

6

Measurements and Testing Components and Circuits

The common problem that you will encounter is with the components, such as, the one mentioned in this chapter, being, either, opened or shorted. Suppose we had a resistor that would not allow any power to go through it; infinite resistance, which would be an open resistor. Now, suppose we had a resistor that allowed, too much, power through, this would be a shorted resistor. Remember, to always make sure your VOM (volt-ohms meter) is zeroed.

Diodes are tested with an ohm and volt meter and the measurement can have slight variation and the type of diode, whether reverse or forward bias, should be known before measuring. Diodes should have zero resistance in one direction and infinite resistance in the other direction. These measurements may not be exact in practical situations; however. The main purposes for diodes in a tv are to provide a stable voltage and alter signals.

You have to remember that the current flows in the cathode and out the anode. The meter has a black and a red wire and plug for each wire The polarity is indicated by the color, the black is for sending electron power, the red is representative of a positive hole that attracts the electrons. When testing the diodes, it is very good practice to unhook one end of the diode, so that, the energy from the ohms meter does not run around a circuit and come back to the other end of the diode. This may make the technician, mistakenly, diagnose the diode. When measuring a zener diode you will use a volt meter and measure the volts across the zener diode. The diode is short or the resistor before the diode is opened, if there is a zero reading. The zener diode is opened if the reading is higher than should be.

Transistors are tested with an ohm and volt meter and the measurements can be different depending on a few exceptions. The transistor base lead can be unplugged from the circuit board to give the most accurate reading. In any case, however, some transistors will give a strange reading because, either, one of the transistors may have a diode or resistor built in, and will give a reading accordingly. Also, it may be a Darlington transistor, in which, it will measure as it is opened.

To measure, either, transistor, you will need to know if it is NPN or PNP. The PNP transistor will measure open if you connect the positive lead at the collector and the negative lead at the base. Vice versa, with an ohm meter, if the positive lead is at the base and the other lead is at he emitter than the transistor is a NPN. Most transistors will have printed on them either NPN or PNP. An open circuit reading from collector to emitter indicates a good transistor; a transistor can be thought of as two diodes.

Capacitors are, either, one of two types, Electrolytic and the ones that are not. The electrolytic capacitors are polarized and are marked with the polarity (+ or – sign). There are many instruments that can be very helpful to measure capacitance, as well. Here is how the capacitance measurements can be measured by the multi-meter (ohms and volt meter) and using the information of the surrounding circuitry.

The reason the circuit can be, critically, important is because, for instance, in a series circuit, we can measure the resistor readings and we can calculate the power at the resistor to find the find $V=IR$ (volts=current*resistance). The capacitance is $V=IXc$ (volts=current*capacitive reactance). The same amount of current is going through the resistor as is going to go through the capacitor; because, it is in a series circuit. The capacitance is then figured by this equation $C= 1/ 2PifXc$. The theoretical value is the value that is written on the capacitor and the measure value is what you measured. If you are not totally familiar with this notation than the table in the appendix can be used as an aid. With these two measurements, an error value can be obtained by subtracting the measured value by the theoretical value and then dividing them by the theoretical value.

Resistors can be measured with the ohms meter. The resistors have color bands on them indicating the value that should be measured. The typical resistor has four bands, where the first two are digits and the third is the multiplier, (the amount of zeros after the first two digits) and the last is the amount of tolerance that can be measured. There is a table in the appendix that shows what colors are represented by the designated number. Measuring resistance is very simple, because the variation of resistor types and the complication of the component is very minimal. The meter ohms selection knob, on the meter, can be adjusted to the proper range, depending on the resistor theoretical value.

Inductors can be measured using a VOM (volt meter); simply, put one lead of the tester on one leg of the inductor and the other lead on the other leg. The inductance measurements will be greatly impacted by the surrounding circuitry, as with, the capacitor. The capacitive reactance (Xc) will decrease when the inductive reactance (XL) increases. Inductors, as with, capacitors and resistors have a color code; a table is in the appendix. Inductors come in many different shapes and sizes and the color code is sometimes obsolete. Measuring inductance is slightly more complicated than some of the other components.

Lt, total inductance, is the sum of the individual inductors in a series circuit. The (ZL) total impedance is the square root of the (RL) coil resistance squared plus the (XL) inductive resistance squared. The (XL) inductive reactance is dependent upon the frequency; XL = 2PifL. Frequency (f) measurements are given a value in Hertz and inductance (L) measurement in Henries. If either, XL or Rl is at least ten times greater than the other than the ZL is equal to the greater value. This, is usually, the situation with XL being greater than RL. To get the theoretical value of the inductors in parallel is calculated using the same method that we use to get total resistance of resistors in parallel. An example, Lt = 1/ (1/ L1+1/ L2+1/L3).

7

Symptom Diagnosis and Troubleshooting

In this chapter, we will break down and troubleshoot the functional block system that we discussed in chapter 1. The Functional block is the very first place to consider when troubleshooting. The problem of various symptoms, from a tv malfunction should, usually, be easily guessed by a technician that has analyzed the circuitry, completely. After reading this entire book you should be able to a have very strong grasp on where any particular trouble lies by any given symptom. By keeping in mind the common tv problem areas, and the functions, and operation of each section of the circuit, you will definitely lead yourself directly to the troubled area. First, you need to experience analyzing the sections of, especially, the most prominent, but all circuits for circuit operation.

The Low-Voltage power supply

The Low-Voltage power supply is the voltage that the tv receives before it is amplified; 120v 60-Hz supply. Problems with the low voltage power supply are usually excessive ripple, no or lowered voltage output. If there is a problem with output than the trouble may be with the regulating transistor or the resistor. First, check all the voltages around the transistor, because this is the common problem; that is, if there is an output at the filter. Often times, there are short circuits that will cause no supply output; likely, if a fuse is blown.

When a picture is shrunken, than there is a problem with very low voltage; voltage supply most likely. The causes of this are open diodes, filter capacitors resistors. When hum is heard in the sound and the picture has horizontal bars or is warped along the side, than, the ripple voltage is probably excessive. This can be checked at the filter capacitor or transistor regulator, the filter capacitor or transistor regulator may need to be changed. Where there is very bright raster and retrace lines in the picture, the tv has emitter and collector leakage. In the high voltage circuit the filter capacitor is the picture tube.

High voltage circuits

High voltage circuits are responsible to keep a picture on the screen, or else, there would be a black screen. The horizontal oscillator, horizontal driver and the horizontal output create the high voltage. Some tvs have high voltage regulators; regulators are made of resistors, and sometimes transistors are used to regulate the power going into the circuit. There are one or more focus voltages of twenty percent of the high voltage that supply the focus control. The symptom of loss of high voltage is loss of raster. High voltage tests can be performed very cautiously and the schematics usually tell the technician what voltage and currents to expect for a normally operating high voltage regulator circuit. High voltage symptoms are very close sometimes to regulator and horizontal output circuits.

When testing for the high voltage malfunction you should first check the low voltage to horizontal output circuit, and then, check the regulator circuits and then, the high voltage circuit. The regulator can be checked by removing it from the circuit and watching to see if the voltage increases. If, it does, the regulator is faulty. High x-rays can ruin the picture tube or cause health problems to the technician if the regulator is left off too long. To say that all these tests have indicated proper operation than we need to, now check the tripler rectifier circuit in the regulator; these must be replaced for the tests to be effective. Assuming, those are fine we now need to check the yoke and fly back. Remember x-rays are produced when the high voltage regulator is disconnected for a long period of time.

Trouble with video circuits can cause a loss of raster. Say, perhaps, that a transistor failed, because of a lack of capacitor coupling, than, the color electron beam on that direct circuit would not function. Another, way to check to see if the video outputs are the problem is to test the cathode of the picture tube and see if the voltage is positive. A short may result in no raster or an open may result in bright retrace lines at the top of the screen; possibly an open capacitor.

Horizontal and Vertical circuits (Sync circuits)

The sync separator passes the signal to the integrator and differentiator circuits after taking it from the video signal. The integrator pulls all the signal pulse into one, in turn, creating the vertical stabilization. The differentiator shapes the pulses, so that, the horizontal sync stays stabilized while the vertical beam is being retraced. It is important to remember the functions of all circuits, so that, troubleshooting will be easier. The loss of or reduced horizontal sweep, improper picture width, loss of horizontal sync, loss of raster and poor horizontal linearity may all be caused by fault in the horizontal circuits; however, not necessarily. Poor linearity, loss of picture height and loss of vertical sync may probably be the cause of a fault in the vertical circuits; however, not necessarily.

Horizontal circuit troubleshooting

The first place we need to start troubleshooting, for Horizontal sync (sweep) Problems, is the oscillator. The oscillator may be bad or is getting a bad signal from the APC. In a case where the oscillator could be running too fast or too slow, the picture shows diagonal lines. These lines point toward the top left of the screen, when the oscillator is too fast and the lines point to the top right of the screen, when the oscillator is running to slow. You will need to know that the oscillator is ran on a time constant where the time of the frequency is based on the capacitors and resistors that control it.

The longer the capacitor takes to charge up the lower the oscillator frequency. Same is with the resistor, when the resistor is smaller the oscillator will be faster. You will need to keep in mind that resistors will increase in value and capacitors will decrease in value in any normal occurrence. The coil of the oscillator is, much like, the capacitor and it can only decrease in value. By varying the adjustment of the horizontal hold control you can test if the oscillator is working or not. If the picture flips from side to side than the frequency is allowed to change and you know that it is working. Now, you know the APC circuit where you need to check next.

Many times, this will be a diode but can be a capacitor or a resistor, as well. Sometimes, but not often, when a set is dead this circuitry could be the trouble. The oscillator is pulling, too much, frequency, and then, the tv is shut down by the automatic shutdown device. You should never use a volt meter on the horizontal circuit after the horizontal circuit begins because horizontal output circuit has, too, high of voltage and will damage your equipment. A scope can be used, however. The picture width can shrink and is many times caused by a transistor, a resistor, capacitor or lowered power supply. To troubleshoot, you can start by checking input voltage of the horizontal circuits, and then, the continue checking the rest of the circuit.

A damper diode is necessary to help prevent oscillation; sometimes the output transistor has the diode inside the transistor. If the damper diode malfunctions you will get a screen with a reduced width or height and /or a no raster. The yoke of the CRT can malfunction and may cause no horizontal sweep, loss of raster or a keystone (narrowing on one side of the screen). A weave on the side of the picture can represent a low voltage power supply, because of ripple getting into the power circuits, leaky filter capacitors or faulty transistor voltage regulators. Tubes and transistors are two different components providing the same functions in the tv. The newer models have transistors and the older one have tubes. The main difference in the two is that, the transistor type has more circuits because the tube is bigger and can handle more power.

Vertical circuit troubleshooting vertical

Frequency is a division of the horizontal 15.734Hz signal and the 262.5 scan lines for a 59.94-Hz vertical frequency. Frequency dividers accomplish this. With the vertical circuit you will be using, at least, the scope, volt meter and ohm meter. The circuitry and troubleshooting is much the same as with the horizontal circuits. Common causes of poor linearity are usually bad capacitors or resistors in the vertical amplifier circuit; and sometimes, fault in the linear feedback circuit. Loss of sync is usually caused by a bad oscillator. Loss of picture height is most usually caused by a faulty output stage, or sometimes, by oscillators and drivers with reduced gain. In all these circumstances, the problem could possibly be components that are aging and adjustments are all that is needed to fix the problem.

A total loss of vertical sweep is when a thick horizontal line is across the middle of the screen. This can be caused by a failure in any of the vertical circuits, which include the driver, oscillator, yoke or output. A shorted yoke, sometimes, causes poor linearity and reduced sweep. A keystone shape appears on the screen in this case. The linearity control circuit can have faulty components and cause undesired feedback, which could cause any of the various problems related to the vertical circuitry functions.

First check, is as with any troubleshooting procedures, the high value components. After checking for these faulty components, such as, leaky capacitors and resistors that changed value check low value components in the circuit that you expect to be the trouble circuit, or the best place to start to find the trouble circuit. As with all the checks, the values for required voltages, resistance and capacitance will be on the schematic. In circuits with an ic for vertical frequency production, instead of an oscillator, you, simply, need to check the input voltage coming from the horizontal output.

Tuners

VHF and UHF tuners are either mechanical or electronical. Mechanical tuners are comprised of a radio frequency amplifier, a mixer and an oscillator. The electronic VHF tuner is comprised of the same components, except the function of the coils that change frequency to change the channel is, instead, done by the change of dc current at the varacter and capacitors. A UHF tuner is without the rf amplifiers and the mixer is a diode; much simpler.

Loss of picture and weak sound reception, snowy pictures and ghosts can all be because of a faulty tuner. Also, i-f, AGC and antenna malfunctions can cause similar problems. A basic process for troubleshooting is to check the fine tuning, check antenna (a color bar generator may be useful here), check tuner inputs AFC or AGC. Then, you could use an analyzer signal or replace the tuner if the problem was not located in the preceding steps. The tuner has many small parts so, be extra careful. A noisy or intermittent channel may be caused by dirty contacts in the tuner. Tuners are located in a metal box to prevent distortion.

I-F circuits

All signal goes through the tuner and is output to the i-f circuits and goes to the if amplifier. Any faults in this circuit will affect picture and audio. The i-f circuit has a video detector circuit that converts the back into video and the sound signal into audio. Traps are tuned circuits that send unwanted signal to ground; an example would be of other interfering channel signals. And, the i-f frequency turns the signal into dc, as discussed earlier.

Where there is complete loss of picture and some audio, where there is complete loss of picture, loss of sync due to clipping, weak picture and normal or weak audio, there may be a problem with the i-f circuit. Our first steps to troubleshooting these, are to, test the contrast control (checking the function of video amplifiers), change the stations or substitute the tuner with another tuner, a tuner box or an analyzer. To test components it is important to check the schematic for the proper signal and test with voltage readings. Signal tracing can be used in this circuit, however may not be efficient. It is quite possible to have an open transistor or anything in that nature.

AGC circuits

When there is too little or too much gain, where the contrast is affected, the AGC circuit may be the cause. White spots and black spots may appear, the picture may bend and or a buzz coming from the tv may be a result of an overload of the AGC circuits. A weak signal, loss of picture and/or no sound may be caused by, too, little gain at the AGC circuit. Other circuits could cause similar problems, such as, the tuner or the i-f amplifier. The first place to troubleshoot is to adjust the AGC or contrast controls to see if they are responding. Next, you would want to create a voltage in the AGC, to use the process of elimination, and then, we will assume it is the tuner or i-f circuit. Next, the tuner is an easy place to substitute a voltage or a tuner itself.

Video circuits and CRT

The video amplifier and CRT circuits and components can be signaled out as a fault, whenever, there is loss of video but no other trouble of any kind. Distorted video, weak video and loss of video are all indication that there is something wrong with these components. There is a possibility that the tv may have four or five video amplifiers. With these problems being the usual, sometimes, there could be sound and picture problem and the problem could be with the video amplifier directly before the audio signal was taken off. Televisions are built differently and the schematics will always be able to tell you, from the block diagram, the exact specifics of where the signals are passing.

When troubleshooting these circuits, it is important to know that there are many types of circuits and circuit designs. In this particular situation, you may need to know if the video amplifier is an emitter follower or a common-base circuit. Also, you will need to be aware if the amplifier is direct-coupled. This knowledge could be helpful in troubleshooting because a short in the collector or emitter of a transistor could shut down that side of the circuit and send the voltage to the base of the direct coupled circuit. This could make you think that the problem is further along in the circuit, after the fault, because there is plenty of juice further along the circuit.

Now, for the video circuits that you will be working with most; the integrated circuits (i-c). The first part of troubleshooting is to, check the input signal to the i-c, then, the second step is to check the output signal and the third step is to check the voltage supply with the i-c removed, to see if the voltage supply is then normal. Then, check the other components around the i-c, just to make sure. Signal injection and signal tracing are used to check for trouble after the i-f circuits. The First place to check is directly after the video detector. If the wave is good than the problem must be with the crt and if, it is not good, check the AGC, by substituting the voltage. If, that is fine, than the i-f circuit or tuner is at fault.

Sound trap circuit, video peaking circuits, contrast control, blanking circuits, automatic picture control and video output circuits could all be other circuits that could cause trouble within the video and crt circuits, however, not as usual. So, now we need to check the crt. The crt is made of a cathode, anode, heater and grids. The test for the signal should be performed on the crt pin. Checking the crt is usually done with an oscilloscope. Checking the crt can be done with a crt tester or by substitution with a test jig. The next tests will be to check for all voltages on the crt.

Color TV circuits

A color bar generator is a very good piece of equipment that will come in handy with color circuitry. Also, the volt meter, scope and ohms meter will be useful. Also, using the Schematics and the block diagram will be very helpful. The color block of the circuitry, in a television, is independent of the rest of the television. A black and white television is the same as the color television, except it does not have the color block; this is not true in later models, however.

In this, situation, if the red and blue are present, but the green is not, than a black screen can not be made. With any problem that is suspected to be in the color circuits, first check, to make sure the circuitry is at fault by turning the color off. If the color is turned off and the picture is good than the trouble is with the color circuitry. Monochrome (one color) can be achieved by adjusting the fine tuning, brightness, contrast and sync controls.

No color sync, Loss of one color, Loss of all color, too much color, all colors wrong and not enough color are all symptoms of the color circuitry at fault. There are six color circuits, burst amp, killer, oscillator, band pass amp, demodulators and video output. For instance, a loss of red will create a blue-green color on the screen. And, with a loss of one color a demodulator is probably the defective section; with a bad capacitor or a coil.

Now, for troubleshooting, always remember to check the controls for minor adjustments. Then, in this case most likely trouble will be with the video output. Loss of all color will require testing with a scope. The burst amp and the band pass amp are the first places to check. Then, the oscillator output and voltage need to be checked. The sync must be 3.58MHz or the demodulators will not work properly. With the loss of all color it is unlikely that the demodulators will all fail at one time which would be necessary for the color to be completely lost. This is the reason why it is the Band pass amplifiers, killer, oscillator and burst amp that would most likely be faulty, in this case.

Too much, color can be caused by amplifier gain through the band pass amplifier, killer and ACC circuitry. Not enough color can be caused by the same as, too, much color. You would need to troubleshoot according to these components and according to the order in the circuit. The trouble with out of sync color is because of an oscillator running on the wrong frequency, or maybe out of phase. If the oscillator checks out good, than the problem components are in the burst amplifier, capacitors, or oscillator circuits.

Audio circuits

Audio circuits in new sets house all the audio circuits in a single i-c with transistors; tubes were used in the old sets. Audio circuits are very simple, the most simple of all the circuits we have discussed. Sound i-f detector, sound i-f amplifier, audio detector and audio amplifiers are the four different circuits in the sound system. Usually, a distorted sound can be caused by a shorted transistor, capacitor or diode. And, usually, reduced but undistorted sound will be a cause of an open capacitor.

When troubleshooting the sound circuits a signal injection of 4.5 MHz FM if should be injected prior to the audio detector. A typical malfunction with audio is no sound. In this case, the audio or speaker may be faulty. The first step to take is to put an ohm meter on the speaker and if a burst noise or a pop occurs than the speaker is good. Another, malfunction with the audio is a when a signal is distorted. In this case, you should use an injection signal and a scope.

8

Advanced Troubleshooting

The technician will be using mainly two items to reference when repairing electronics, manufacture's manual and the schematic. The manufacturer's manual has information of particular and general information relating to the circuitry of the tv; made by the manufacturer. For instance, instructions for control adjustments, servicing the power supply etc…The schematic will have all of the circuitry mapped out and most often all the measurements and details. It is certainly possible to fix the television without a schematic, however, it could get confusing; especially for a beginner. A schematic, will also show some spots in the circuit where testing would probably be the most suitable to ease the troubleshooting and repair process.

Starting troubleshooting from the functional block diagram will be most practical. The next step, most likely, will be to measure input and output waveforms to narrow down the part of the circuit where the fault is at. With this approach it is necessary to remember that the horizontal circuit could be a cause of the whole set being dead. The tv is micro controlled and this is where the problem will most likely be. The large chip will control the power supply, sync, tuner and closed caption data.

Now, let's discuss some of the points to remember of each block of the tv. The schematic does not have any recommended test location for the tuner. This is because the tuner is very delicate circuitry. However, a signal generator can be used to test to make sure that the i-f circuit is working properly. The signal would be injected and the set would operate properly if the tuner was bad. The composite video waveform can also be tested by using the oscilloscope at the resistor at the input of the circuit.

A composite video signal can be injected into the horizontal circuit, vertical circuits, chroma-luminance circuitry and into the audio to test for proper function. This is a way to test sweep trouble and can lead you to other parts of the surrounding circuitry. The pulse width modulation circuitry can be determined by using an oscilloscope. If the sweep is shut down it can be because the width modulation circuitry is supplying too much voltage. When testing the chrome-luminance a color bar generator will be useful as follows. The first part that may need to be tested is the input to the i-c circuit, then, the input, and then, the output of the color drivers (the three transistors on the video output board) will need to be checked. After this, and there is no faulty components, check the surrounding resistors.

9

Soldiering Techniques

Always use safety glasses while soldering. The surface that is to be soldered needs to be very clean. Rosin flux is needed to help keep the surface clean; flux is usually in the solder. Be very careful not to overheat the circuit board so that the foil does not peel off. Cold soldier joints are the biggest of problems for beginners of soldiering. The cold soldier joint is when soldier is not totally adhering to the board or the leg of a component etc…To prevent from getting cold soldier joint you should practice some until you get used to the way that the soldier melts from the blend of heat that is given by the iron and movement of the iron.

When placing a component in the hole of the circuit board to be soldered the bend of the component should only be slight, at a ninety degree angle and not, too, sharp. Also, when soldering components on the circuit board the leads of the component can be bent some on the backside of the board, so that, the part does not fall out until it is soldered. Then, the leads are clipped off. There are tools that will allow you to easily and cleanly remove solder from a circuit board. Solder-wick, vacuum tube and the de-soldering tool are very handy for these purposes.

Epilogue

There is very much to learn in the field of electronics, especially, if you are studying the whole field. However, this book will give you the basics you need to start or improve at being skilled in the profession of electronics troubleshooting and repair.

APPENDIX A

This section of the appendix includes only some of the fundamental circuit related components, semiconductors and switches.

Circuit Symbol

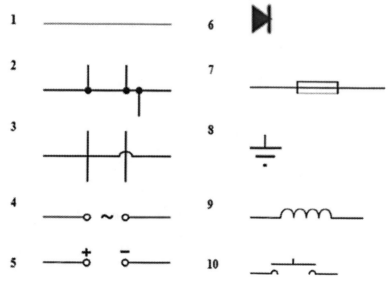

1. wire

2. wires joined

3. wires not joined

4. a-c power supply

5. d-c power supply

6. diode

7. fuse

8. ground

9. inductor

10. push switch

11

12

13

14

15

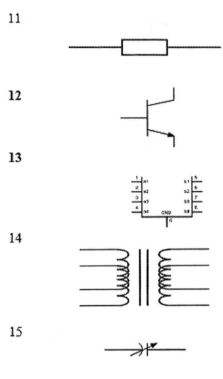

11. resistor

12. transistor NPN

13. base of ic

14. transformer

15. variable capacitor

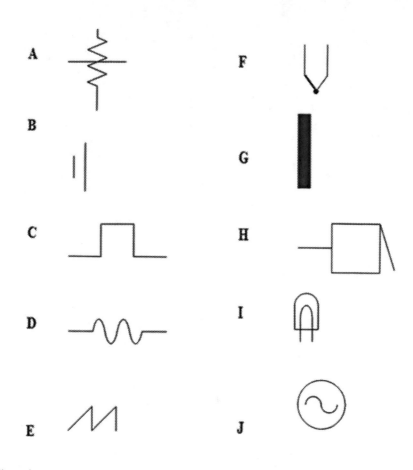

A. attenuator

B. accumulator

C. pulse

D. alternating pulse

E. sawtooth waveform

F. Thermoconductor

G. ferrite core

H. buzzer

I. lamp

J. oscillator

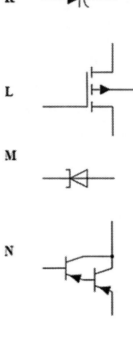

K. varactor
L. mosfet
M. backward diode
N. darlington
O. surge protector

APPENDIX B

This section of the appendix is much like the first. This appendix is to show how similar symbols are related in shape compared to other symbols in there close family.

Electrical circuit components

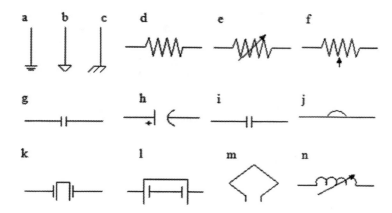

a) Ground
b) Common Ground
c) Chassis Ground
d) Resistor
e) Variable Resistor
f) Adjustable Resistor
g) Non-Polarized Capacitor
h) Electrolytic Capacitor
i) Contact
g) Circuit Breaker
k) Crystal
l) Delay
m) Antenna
n) Variable Inductor

APPENDIX C

This section of the appendix shows some common semi conductor symbols that are closely related in symbol shape.

Semiconductor devices

a) NPN Transistor
b) PNP Transistor
c) Triode PNPN Switch
d) Bidirectional Breakdown Diode PNP
e) Breakdown Diode NPN
f) Bidirect, Breakdown Diode NPN
g) Bidirectional Breakdown Diode PNP
h) Zener Diode
i) Rectifier
g) Varactor
k) Tunnel Diode
l) N Base Unijunction Transistor
m) P Unijunction Transistor

APPENDIX D

Resistor types and color coding

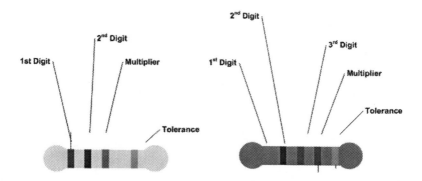

Resistor Color Code Table

Color	1st Digit	2nd Digit	3rd Digit	Multiplier	Tolerance
Black	0	0	0	1Ω	1%
Brown	1	1	1	10Ω	2%
Red	2	2	2	100Ω	
Orange	3	3	3	1,000Ω	
Yellow	4	4	4	10,000Ω	
Green	5	5	5	100,000Ω	.5%
Blue	6	6	6	1,000,000Ω	.25%
Violet	7	7	7	10,000,000Ω	.10%
Grey	8	8	8	100,000,000Ω	.05%
White	9	9	9	1,000,000,000Ω	
Gold				.1Ω	5%
Silver				.01Ω	10%
No Color					20%

APPENDIX E

Capacitor types and color coding

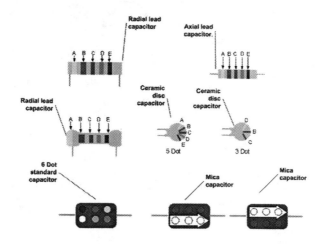

Ceramic Capacitor Color Code Table

Color	1st Digit	2nd Digit	Multiplier	Tolerance (+/-) over 10pF	Tolerance (+/-) under 10pF	Temperature coefficient
Black	0	0	1	20%	2.0pF	0
Brown	1	1	10	1%		-30
Red	2	2	100	2%		-80
Orange	3	3	1,000			-150
Yellow	4	4	10,000			-220
Green	5	5	100,000	5%	0.5pF	-330
Blue	6	6	1,000,000			-470
Violet	7	7	10,000,000			-750
Grey	8	8	.01		.25pF	+30
White	9	9	.1	10%	1.0pF	+120 to -750(BA) +500 to -30 (JAN)
Gold						Bypass or Coupling
Silver						+100 (JAN)

6-Dot Standard Capacitor Color Code for RMA, JAN, AWS

Title	Color	1st Digit	2nd Digit	Multiplier	Tolerance (+/-) Percent
JAN,MICA	Black	0	0	1	20%
	Brown	1	1	10	1%
	Red	2	2	100	2%
	Orange	3	3	1,000	3%
	Yellow	4	4	10,000	4%
	Green	5	5	100,000	5%
	Blue	6	6	1,000,000	6%
	Violet	7	7	10,000,000	7%
	Grey	8	8	100,000,000	8%
BA, MICA	White	9	9	1,000,000,000	9%
	Gold			.1	
Molded paper	Silver			.01	10%
	No Color			.01	20%

MICA Capacitor Color Code

Color	1st Digit	2nd Digit	Multiplier	Tolerance (+/-)Percent	Voltage Rating
Black	0	0	1	1%	100
Brown	1	1	10	2%	200
Red	2	2	100	3%	300
Orange	3	3	1,000	4%	400
Yellow	4	4	10,000	5%	500
Green	5	5	100,000	6%	600
Blue	6	6	1,000,000	7%	700
Violet	7	7	10,000,000	8%	800
Grey	8	8	100,000,000	9%	900
White	9	9	1,000,000,000		1000
Gold	0	0	.1	10%	2000
Silver	1	1	.01	20%	

APPENDIX F

Scientific notation

Prefixes for Exponents

Prefix	Symbol	Scientific Notation	Form
tera	T	10^{12}	1,000,000,000,000
giga	G	10^9	1,000,000,000
mega	M	10^6	1,000,000
kilo	k	10^3	1,000
hector	h	10^2	100
deka	dk or da	10^1	10
deci	d	10^{-1}	0.1
centi	c	10^{-2}	0.01
milli	m	10^{-3}	0.001
micro	u	10^{-6}	0.000,000,001
nano	n	10^{-9}	0.000,000,000,001
pico	p	10^{-12}	

Afterward

Through the chapters we have discussed the many basics of electronics grouped into their perspective sections. The information can be a lot to memorize but understanding the independent and interdependent relationships within the entire block and entire television can make troubleshooting and diagnosis symptom questions easier to answer. The most descriptive and complete overview was given to provide a very good and very quick foundation for future learning in the subject.

Conclusion

The chapters were written, so that, the reader could get the details of the common jobs performed by the consumer electronics technician; particularly, the television technician. By knowing how the circuits and components function together, troubleshooting is much easier. Using the ohms law and the Kirchhoff's law one can very easily understand the basics of how each circuit was designed and is intended to function. Realistically, for the consumer electronics repair technician, the repairs are, not so, abundantly various that we cannot draw up a guide to make the job less work and more productive. The chores of tracking down problem circuitry have been made very simplistic with the layout of troubleshooting in this book. The repair technician is very successful when concentrating on the technique and proper repair procedures.

About the Author

I have achieved a formal education for Advanced Technology in Consumer Electronics Troubleshooting I and II. Also, I have studied and researched consumer electronics repair and gained very much knowledge in the subject. Aside, from text book study, I learned through actual experiences with repair jobs. I was, also, fortunate to have a friend that owned his own business of consumer electronics repair.

I worked with him on many road trips and repairs in the shop. I started out wondering much about where, how, why and what any particular circuit or component (s) were involved in faulty equipment. I asked a lot of questions at first about troubled circuit malfunction and/or about the makeup of particular circuits and the purpose of each device. Thinking through my experiences, I decided to put together a simple and in depth book.

Cleveland Institute of Electronics

Witness that

Doug Kinney

has satisfactorily completed all requirements for the course

Electronics Technology and Advanced Troubleshooting I & II

and as proof of proficiency is hereby awarded this

Diploma

In Testimony Whereof, the Institute has caused this
Diploma to be signed by the authorized officials and
the seal affixed at Cleveland, Ohio,
on this 22nd day of November, 2002

President

Vice President

978-0-595-41171-9
0-595-41171-1

Printed in the United States
91004LV00009B/73/A